Kuiper Belt

Lily Erlic

DEEP in SPACE

www.av2books.com

Go to www.av2books.com, and enter this book's unique code.

BOOK CODE

AVS73585

AV^2 by Weigl brings you media enhanced books that support active learning.

AV^2 provides enriched content that supplements and complements this book. Weigl's AV^2 books strive to create inspired learning and engage young minds in a total learning experience.

Your AV^2 Media Enhanced books come alive with...

Audio
Listen to sections of the book read aloud.

Video
Watch informative video clips.

Embedded Weblinks
Gain additional information for research.

Try This!
Complete activities and hands-on experiments.

Key Words
Study vocabulary, and complete a matching word activity.

Quizzes
Test your knowledge.

Slideshow
View images and captions, and prepare a presentation.

... and much, much more!

Published by AV^2 by Weigl
350 5th Avenue, 59th Floor
New York, NY 10118
Website: www.av2books.com

Library of Congress Control Number: 2019941862

ISBN 978-1-7911-0962-2 (hardcover)
ISBN 978-1-7911-0963-9 (softcover)
ISBN 978-1-7911-0964-6 (multi-user eBook)

Printed in Guangzhou, China
1 2 3 4 5 6 7 8 9 0 23 22 21 20 19

052019
102918

Project Coordinator: John Willis
Designer: Terry Paulhus

Every reasonable effort has been made to trace ownership and to obtain permission to reprint copyright material. The publishers would be pleased to have any errors or omissions brought to their attention so that they may be corrected in subsequent printings.

Weigl acknowledges Alamy, Getty Images, iStock, and NASA as its primary image suppliers for this title.

CONTENTS

Kuiper Belt

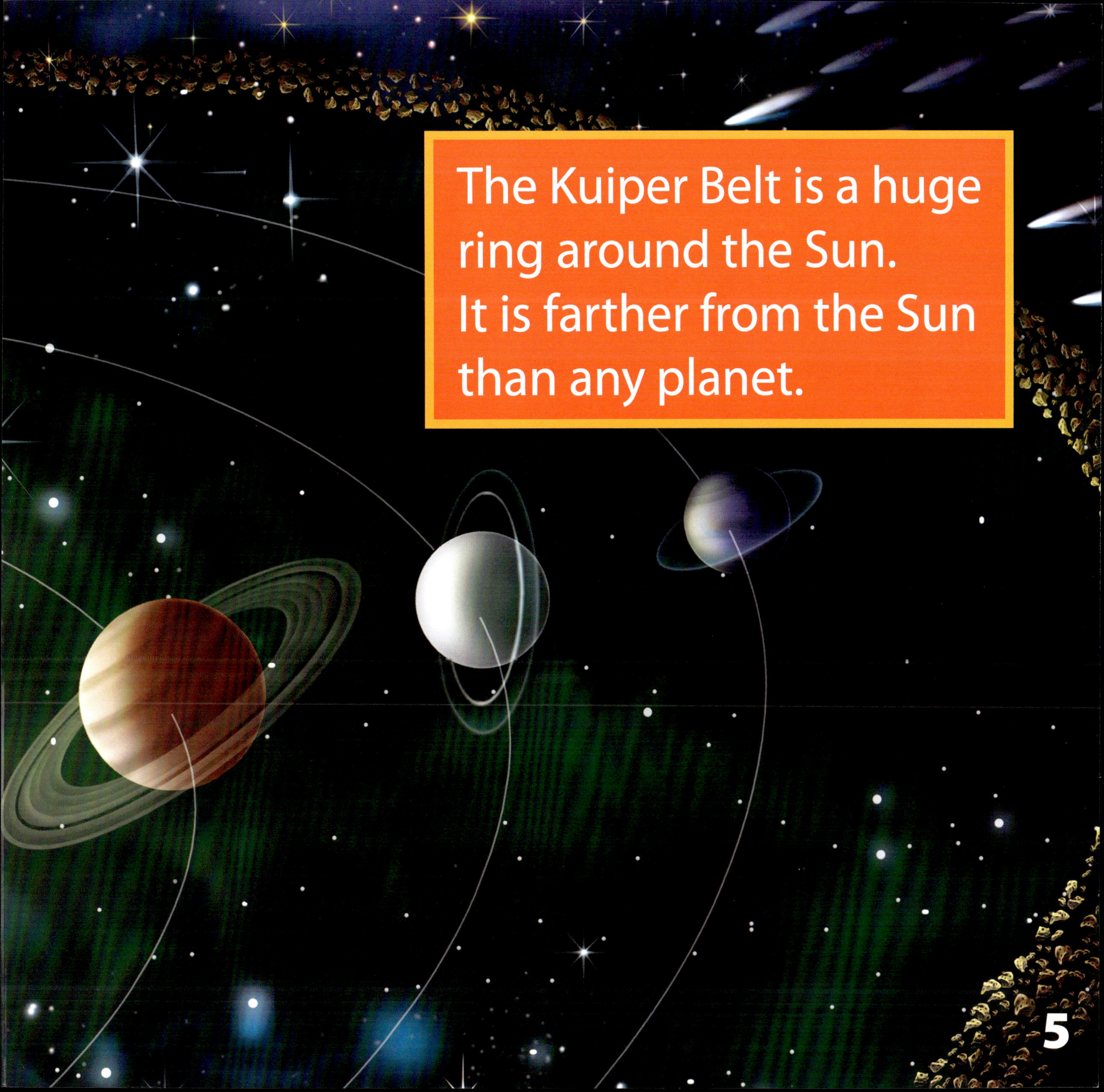

The Kuiper Belt is a huge ring around the Sun. It is farther from the Sun than any planet.

The Kuiper Belt is inside a giant bubble of gas and ice. The bubble is called the Oort Cloud.

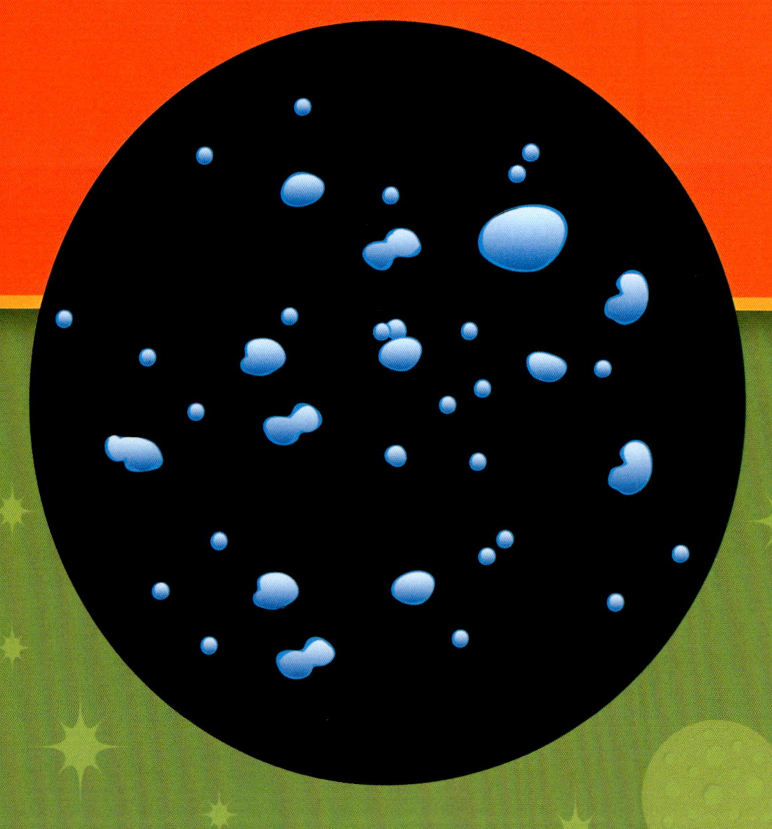

Kuiper Belt
Oort Cloud

The Kuiper Belt is made of ice, rocks, and dwarf planets. Dwarf planets are smaller than planets.

Pluto is in the Kuiper Belt.
People used to call it a planet.
Then, they found other small objects.
People called them all "dwarf planets."

Pluto is not the only dwarf planet in the Kuiper Belt.

Makemake is also a dwarf planet. It is known for its red color.

Haumea looks like a big egg. It is the only known dwarf planet with a ring.

Comets are balls of rock and ice. People think some comets come from the Kuiper Belt.

Comets have tails made of gas. The gas glows when it gets close to the Sun.

Scientists sent a spacecraft called *New Horizons* to study the Kuiper Belt. It passed by Pluto in 2015.

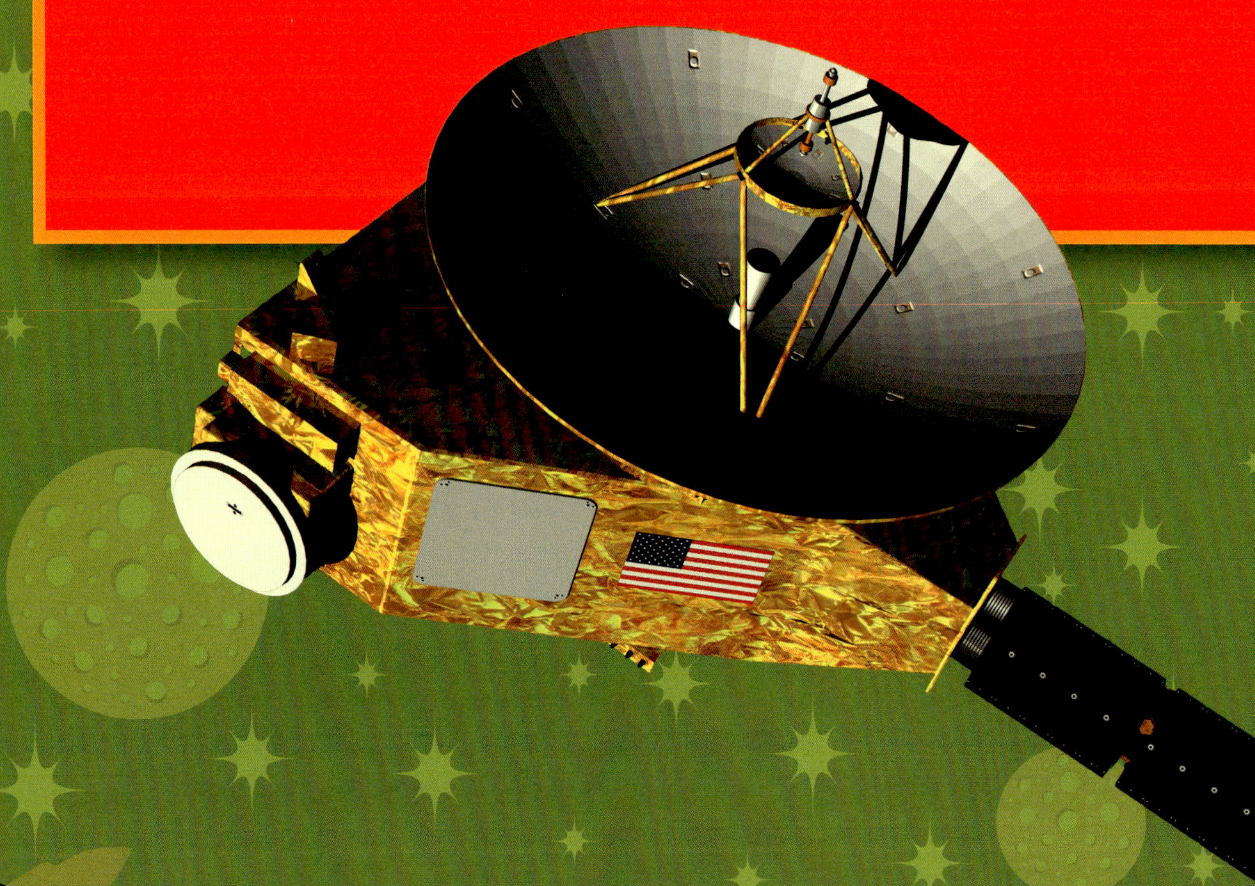

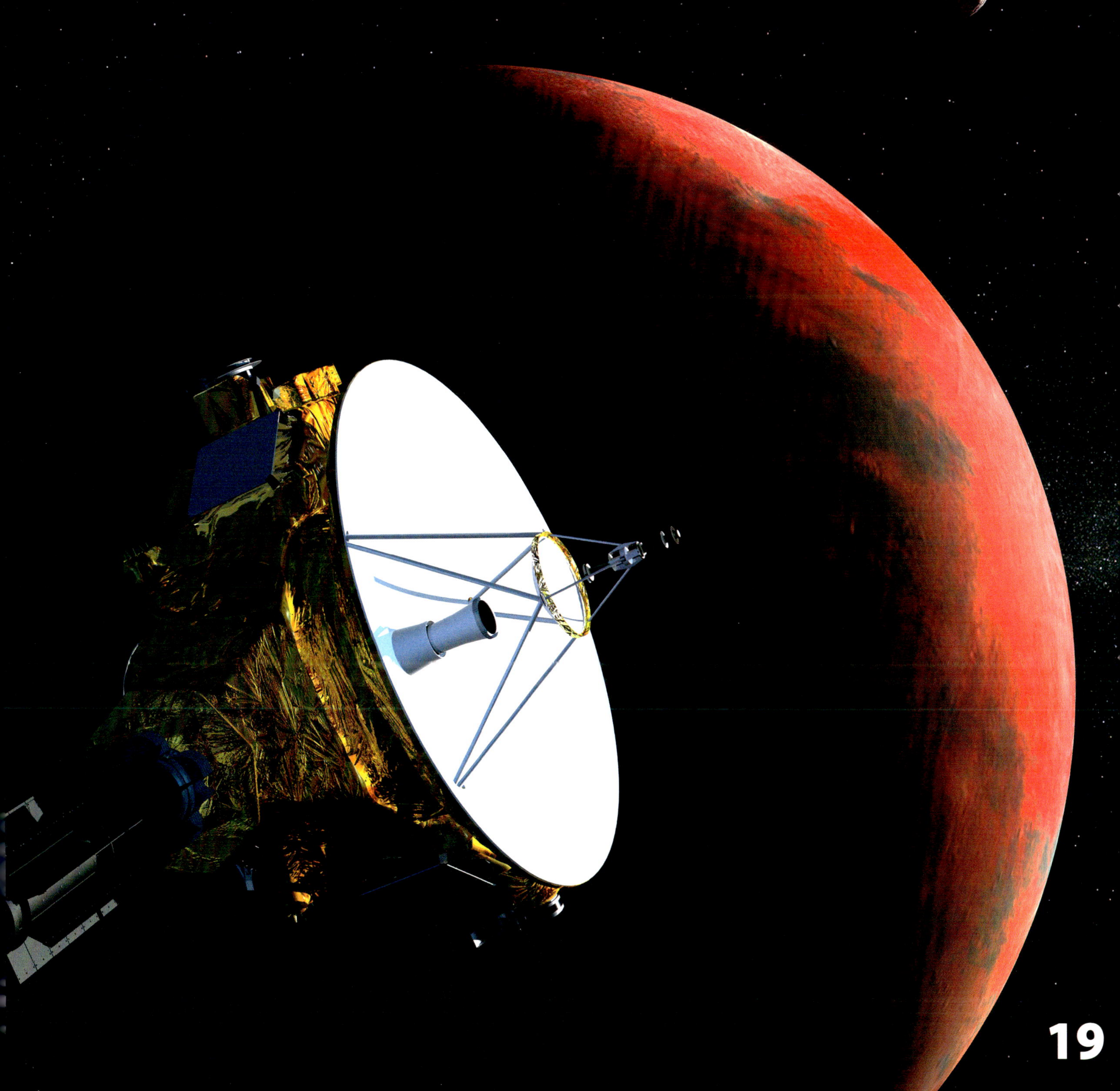

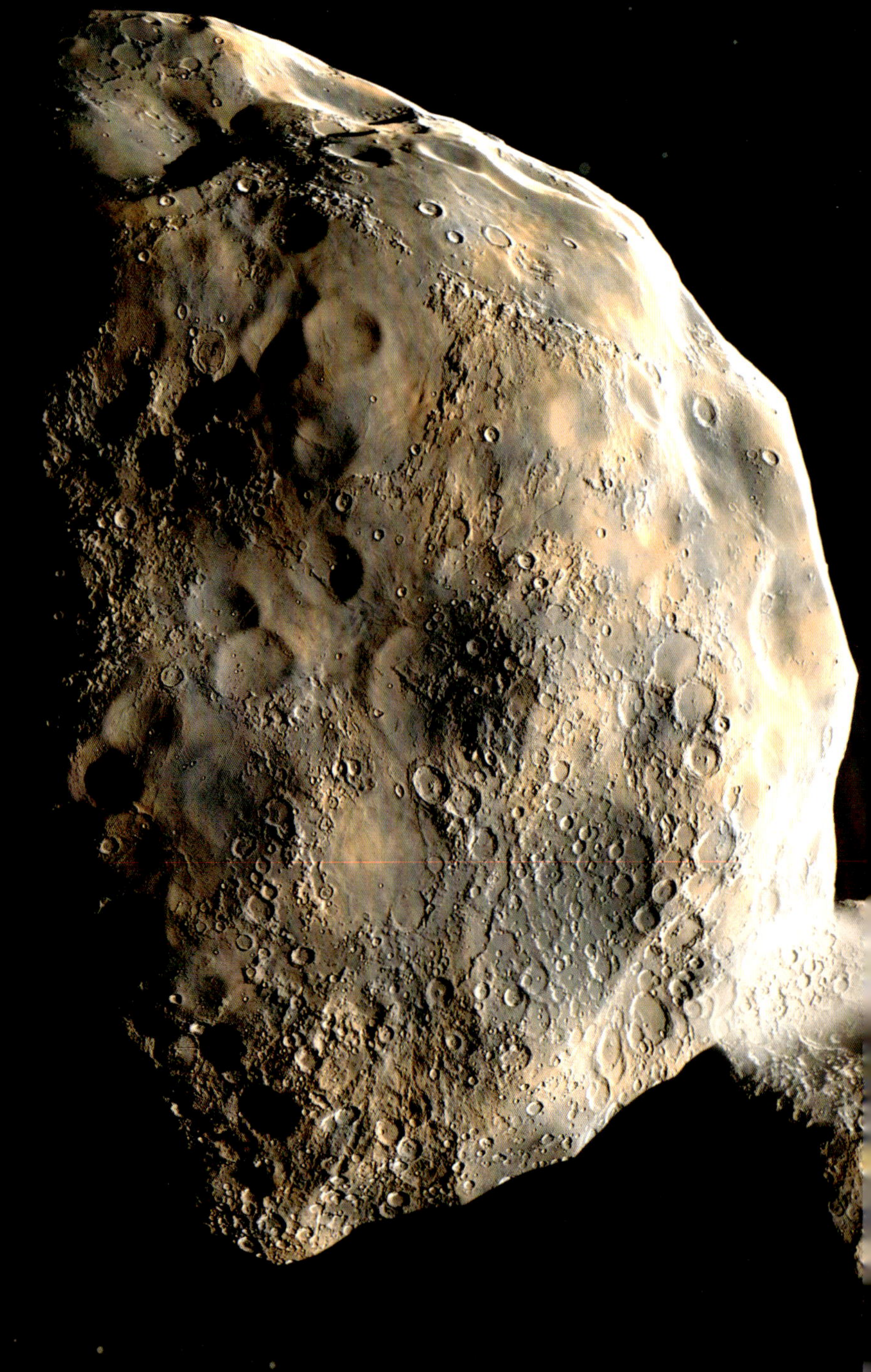

New Horizons found a new object in the Kuiper Belt. It is the farthest object a spacecraft has ever visited.

KUIPER BELT FACTS

These pages provide detailed information that expands on the interesting facts found in the book. They are intended to be used by adults as a learning support to help young readers round out their knowledge of each object or event featured in the *Deep in Space* series.

Pages 4–5

The Kuiper Belt is a huge ring around the Sun. The inner edge of the Kuiper Belt is around 2.8 billion miles (4.5 billion kilometers) from the Sun. It was named after astronomer Gerard Kuiper. Other astronomers have discovered many space objects such as comets, dwarf planets, rocks, ice, and moons in the Kuiper Belt. Like Earth, the objects in the Kuiper Belt orbit the Sun.

Pages 6–7

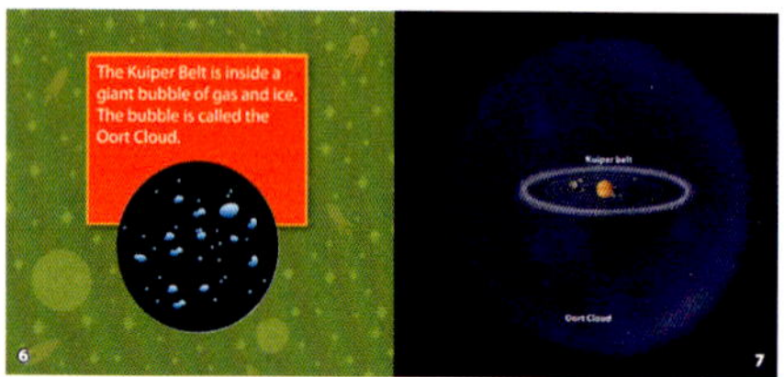

The Kuiper Belt is inside a giant bubble of gas and ice. The Oort Cloud surrounds the solar system, including the Kuiper Belt. The cloud is trillions of miles wide. Scientists think it could contain about 2 trillion objects. Long-period comets come from the Oort Cloud. They are brighter and longer in shape than short-period comets. Scientists think the well-known Halley's Comet came from the Oort Cloud.

Pages 8–9

The Kuiper Belt is made of ice, rocks, and dwarf planets. There are four dwarf planets in the Kuiper Belt. Like regular planets, dwarf planets orbit the Sun. Scientists have found a total of five dwarf planets in the solar system. The only one that is not in the Kuiper Belt is called Ceres. It is in an asteroid belt between Mars and Jupiter.

Pages 10–11

Pluto is in the Kuiper Belt. Pluto is one of the brightest objects in the Kuiper Belt. It is 1,473 miles (2,371 km) wide. Pluto has five moons. The National Aeronautics and Space Administration (NASA) sent a spacecraft to study Pluto in 2006. Scientists discovered new things about the dwarf planet. They saw mountains, a heart-shaped icy plain, and frozen red gas.

Pages 12–13

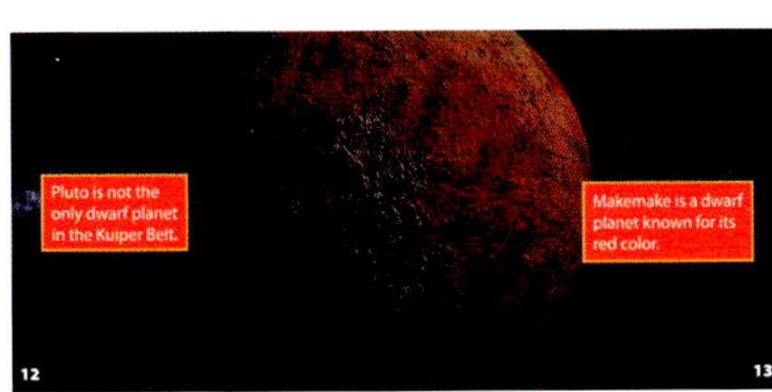

Pluto is not the only dwarf planet in the Kuiper Belt. Makemake is the third-largest dwarf planet. Astronomers discovered it in 2005. Makemake has one moon. It is dark and very cold on Makemake. Its surface is covered by frozen gases. When the gases reflect light from the Sun, they give Makemake a reddish glow. From Earth, Makemake appears to be the second-brightest object in the Kuiper Belt.

Pages 14–15

Haumea looks like a big egg. It is almost the same size as Pluto. Haumea is tiny compared to Earth. If Earth were the size of a nickel, then Haumea would be about the size of a sesame seed. Haumea spins quickly. It is the fastest-spinning dwarf planet in the solar system. The light of the Sun takes six hours to reach Haumea.

Pages 16–17

Comets are balls of rock and ice. Comets that come from the Kuiper Belt are called short-period comets. These comets take less than 200 years to go around the Sun. They are made of ice and dust. There are millions of comets in the Kuiper Belt. From Earth, comets appear to have a glowing tail. Comets actually have two unique tails, one made of gas and the other made of dust.

Pages 18–19

Scientists sent a spacecraft called *New Horizons* to study the Kuiper Belt. It travels approximately 36,000 miles (58,000 km) per hour. It took *New Horizons* about nine and a half years to arrive near Pluto. The spacecraft has cameras that record pictures of icy objects, planets, and moons in the Kuiper Belt. It can take months for *New Horizons* to send a photograph back to scientists on Earth.

Pages 20–21

***New Horizons* found a new object in the Kuiper Belt.** Scientists used a space telescope to take pictures of Kuiper Belt objects, or KBOs. One was a KBO made of two icy objects attached to each other. It looks like a lumpy snowman. Scientists call it *Ultima Thule*, which means "a distant, unknown region." Ultima Thule was one of the first KBOs to be explored by a spacecraft.

KEY WORDS

Research has shown that as much as 65 percent of all written material published in English is made up of 300 words. These 300 words cannot be taught using pictures or learned by sounding them out. They must be recognized by sight. This book contains 43 common sight words to help young readers improve their reading fluency and comprehension. This book also teaches young readers several important content words, such as proper nouns. These words are paired with pictures to aid in learning and improve understanding.

Page	Sight Words First Appearance
5	a, any, around, from, is, it, than, the
6	and, of
9	are, made
10	all, call, found, in, other, people, small, them, then, they, to
12	not, only
13	also, for, its
14	big, like, looks, with
16	come, some, think
17	close, gets, have, when
18	by, study
21	has, new

Page	Content Words First Appearance
5	Kuiper Belt, planet, ring, Sun
6	bubble, gas, ice, Oort cloud
9	dwarf planets, rocks
10	objects, Pluto
13	color, Makemake
14	egg, Haumea
16	comets
17	tails
18	*New Horizons*, scientists, spacecraft

Check out av2books.com for activities, videos, audio clips, and more!

1. **Go to av2books.com**
2. **Enter book code** AVS73585
3. **Explore your book!**

www.av2books.com